Carnivorous plants

Carnivorous plants

A Wisley handbook

Paul Temple

Cassell

The Royal Horticultural Society

Cassell Educational Limited
Artillery House
Artillery Row
London, SW1P 1RT
for the Royal Horticultural Society

First published 1988
Second impression 1989

British Library Cataloguing in Publication Data

Temple, Paul, *1951*
    Carnivorous plants. – (A Wisley Handbook)
    1. Insectivorous plants – Great Britain
    I. Title   II. Royal Horticultural Society
    III. Series
    635.   SB432.7

    ISBN 0-304-31145-6

Photographs by Sandra Banfield, The Harry Smith
Collection and Michael Warren
Design by Lesley Stewart
Phototypesetting by Chapterhouse, Formby
Printed in Hong Kong by Wing King Tong Co. Ltd.

Contents

Above: the delightful flowers of the butterworts – *Pinguicula moranensis caudata* on the left and two hybrids on the right

Below: the Venus fly trap, *Dionaea muscipula*; in good light the traps often become deep red

Introduction

'Carnivorous plants' is a convenient term for a group of plants, some closely related and others not, which are linked by a single common character – that of trapping and digesting living organisms in order to supplement the food provided by other means. Although often causing raised eyebrows or feelings of distaste, these peculiar plants have always had a small but enthusiastic following and have indeed enjoyed two periods of wider popularity, first during the late Victorian age and secondly in the 1970s, when a revival of interest began that is still with us.

Despite the numerous myths surrounding them, carnivorous plants are hardly the man-eating fiends so many people seem to have heard of. Even the largest plants generally restrict their diets to insects and other creepy-crawlies; and while the tropical pitcher plants or Nepenthes can grow into long scrambling vines, the vast majority of carnivorous plants will be no more than about 3 ft (1 m) tall.

Common belief also suggests, incorrectly, that all carnivorous plants have traps which snap shut on their unsuspecting prey. The truth is that many of them do not move at all, but instead rely on insects falling into traps which are too slippery to allow escape. The misunderstanding has arisen from the well known Venus fly trap, Dionaea, which closes so dramatically on its victim. However, this is merely one of several types of trap and the pitcher trap, a design found in Sarracenia and Nepenthes, does not make any movement.

Yet another misconception is that carnivorous plants are very difficult to grow and not easily obtained. In fact, their cultivation is generally straightforward and they require less attention than many non-carnivorous plants. For example, there is no need to worry about when and how much fertilizer to apply: carnivorous plants catch their own! There are certainly a few guidelines that will help the gardener produce the best results, but these are simple to remember and follow. Failure is almost always due to incorrect advice or to the purchase of sickly plants through inexperience. Healthy plants are now readily available from specialist nurseries and also from an increasing number of garden centres, both of which supply the popular species and forms at inexpensive prices.

Finally, no introduction to the subject would be complete without mentioning that carnivorous plants can be grown in the

Drosera rotundifolia, a hardy northern sundew, growing wild in sphagnum bog on a Welsh mountain

garden. Few people would even consider the possibility of growing them outdoors and yet there are several carnivorous plants that are either fully hardy or half-hardy. These are worthy of a place in any garden where the owner takes a pride in achieving variety and interest.

When the myths are exploded, it can be seen that carnivorous plants present a marvellous opportunity for gardeners to enhance their plant collections, whether they appreciate the beautiful, the bizarre or simply the chance to try something new. This book is intended to give practical advice, which will enable anyone to grow carnivorous plants, and to provide an introduction to the most important and worthwhile genera, together with a selection of recommended plants.

What is a carnivorous plant?

Carnivorous plants share only one feature – the ability to trap other living organisms to be used as a source of food. The plants that fall within this definition need not be botanically related and therefore they show a wide variety of forms, including several distinct types of trapping mechanism. Remarkable as it may seem, almost all traps are highly modified leaves.

To be accepted as carnivorous, a plant must not only have a mechanism for trapping, but the trap must also be specifically designed to extract food from the prey. Associated with the trap, there should be cells that can assist with either digesting the prey or absorbing the resulting food. Many carnivorous plants have both types of cell.

It is rare for plants to be carnivorous. In addition to several fungi and a rather strange example represented by a species of oak gall, there are only eight known families of flowering plant which include or entirely consist of carnivorous species (see table 1). Even so, well over 1,600 species and hybrids have so far been named. New species are still being identified and it is likely that there are more awaiting discovery.

Table 1: carnivorous plant families and genera

Family	Genus	Popular name
Bromeliaceae	*Brocchinia*	
Byblidaceae	*Byblis*	rainbow plant
Cephalotaceae	*Cephalotus*	Albany pitcher plant
Droseraceae	*Aldrovanda*	waterwheel plant
	Dionaea	Venus fly trap
	Drosera	sundew
	Drosophyllum	Portuguese sundew
Dioncophyleaceae	*Triphyophyllum*	
Lentibulariaceae	*Genlisia*	
	Pinguicula	butterwort
	Utricularia	bladderwort
	Polypompholyx	fairy apron
Sarraceniaceae	*Darlingtonia*	cobra lily
	Heliamphora	sun pitcher
	Sarracenia	North American pitcher
Nepenthaceae	*Nepenthes*	monkey cup

WHY FOOD IS TRAPPED

Carnivorous plants are found throughout the world and in many different habitats – from the acid peat bogs of Britain, where *Drosera* and *Pinguicula* grow, to the tropical jungles of southeast Asia, inhabited by the *Nepenthes*, and the flooded savannahs of Africa, which provide a home for *Utricularia*.

Many carnivorous plants require permanently wet conditions, as this might suggest, although others live in very dry sandy soils. However, the characteristic common to all their habitats is the lack of nutrients in the soil. By lessening their dependence on the soil for nourishment and by supplementing their diet with food that literally walks or flies in, the carnivores are able to survive where few other plants can. This visiting animal life is an important source of nitrogen – the nutrient which is most easily lost from the soil and which is always rare in soils colonized by carnivorous plants – and nitrogen helps to increase leaf growth and to improve or make possible both flowering and seed production.

TYPES OF TRAP

Traps can be divided initially into active and passive types. Active traps employ movement in the process of trapping and digesting their prey, *Dionaea* being the best known example. The mechanism is reminiscent of an old-fashioned game trap, with two sections hinged open and sprung, waiting to be triggered by the unsuspecting victim. *Pinguicula* use no movement to catch food with their fly-paper type leaves, but because they roll their leaf margins immediately after trapping to assist digestion they are also classified as active. Similarly, *Drosera* bend dew-dropped tentacles covering the leaf surface towards the prey, which is stuck to the adhesive droplets, and sometimes bend the entire leaf to surround their food. While they are not as fast as most other active traps, a very few species can bend their leaves through 180° in less than 60 seconds.

Passive traps such as *Sarracenia* utilize no movement either in trapping or during digestion. They simply expect the prey to move towards the plant and become ensnared, by falling into a pit or being stuck with a gluey substance.

Traps can also be distinguished by their form. The largest group, known as pitfalls, include all the pitchers. These consist of a tube into which a creature will eventually tumble and are often enhanced with nectaries or colouring to attract the prey, as well as hairs or waxy cells to assist trapping. *Sarracenia* is a good example of a pitcher which may employ the full range of available inducements. Fly-paper traps, like those of *Drosera*, use sticky mucus to glue the prey to the leaf blade. The steel-trap of Dionaea snaps shut on any creature unfortunate enough to trigger the sensitive hairs on the trap surface. Mousetraps, such as those of Utricularia, work in a very similar fashion, but suck in the prey in response to being triggered, operating in very wet conditions or under water. Table 2 identifies the types of trap found in the most popular genera.

Table 2: carnivorous plants and their trap types

Genus	Passive trap type	Active trap type
Cephalotus	pitfall	
Dionaea		steel-trap
Drosera		fly-paper
Pinguicula		fly-paper
Utricularia		mousetrap
Darlingtonia	pitfall	
Sarracenia	pitfall	
Nepenthes	pitfall	

DISTRIBUTION

Carnivorous plants are much more widespread than most people assume. They occur in the extreme north of the Soviet Union and Canada and far south into Tierra del Fuego at the tip of South America, and in all climates from temperate to tropical. Only the islands of the Pacific Ocean seem to be devoid of any carnivorous plant species.

Some species have a very wide distribution. For instance, the most common British carnivorous plant, Drosera rotundifolia, is found in Asia, the USSR, Europe and North America. Others are restricted to extremely limited areas, like the rare genus Heliamphora, which is confined to a tiny isolated location in the Roraima mountains of Venezuela.

Cultivation

This chapter gives general information on the cultivation requirements of carnivorous plants; for further details, see the individual descriptions on pp.32–55.

COMPOSTS

Commercially available composts are not recommended for carnivorous plants, but it is easy and cheap to mix suitable composts at home and does not require precise measurement or many ingredients. Care must be taken to ensure that the mixture is made up in the correct proportions, as carnivorous plants vary in their preferences, from those needing free drainage to those wanting permanently wet bog. Most plants fall into one of four categories, for which appropriate mixtures are shown in table 3.

Table 3: types of compost

Compost type	Ingredients
A	Equal parts by volume of moss peat and washed sharp sand
B	Equal parts by volume of moss peat, washed sharp sand and perlite or vermiculite
C	Live sphagnum moss with some charcoal
D	Pure live sphagnum moss

All these composts are without nutrients. This is important, because carnivorous plants always grow in infertile soils or water. Addition of fertilizer to the compost may cause root damage, malformation of the trapping leaves or death of the plant and therefore should not be attempted.

Opposite above: *Drosera capensis*, one of the easiest sundews to grow and ideal for a windowsill

Below: the beautifully marked *Sarracenia leucophylla* will enhance any collection

Sharp sand is used to give the compost an open texture. The grains should measure 1/16 to 1/8 in. (1.6–3 mm) and should be washed first to remove as many chemicals as possible. To do this, partly fill a bucket or bowl with the sand and add ordinary tap water. Agitate the sand with your hands or a stick and then pour off the water, repeating until the water runs clear. For larger amounts, fill one third of a deep container with the sand and flush it out with water from a garden hose, pushing this well into the sand to stir it up. If the sand is not washed, there is a tendency for chemical impurities to leach on to the soil surface and leave ugly orange-brown deposits, which severely detract from the beauty of the plants. The sharp sand should be acid or neutral, as not all carnivorous plants are lime-tolerant. River sand of the same grain size, when available, is the best kind to use.

Sphagnum moss is less common as a growing medium than is often supposed. It retains high volumes of water and can damage some plants if left around the plant neck, particularly in winter; in the right conditions, it also grows very quickly and may smother small plants. Therefore, it should only be used where specifically recommended. Sphagnum itself is now rare in the wild, many sites being protected by conservation orders. However, it is easy to buy or grow. Garden centres usually offer sphagnum for sale in the spring and summer months and any spare moss can be stored moist in a plastic bag kept in a shady place. For longer storage, pour off excess water and refrigerate or freeze it. To grow even a small amount of sphagnum, a large shallow container is necessary and plenty of growing space. The shredded sphagnum is scattered on to the container and rain or distilled water added (not hard tap water as it is intolerant of lime). If placed in the light and kept wet, the moss will grow.

CONTAINERS

Plastic containers are normally used for carnivorous plants and have several benefits, being cheap, easy to sterilize, light and less fragile than clay. In the very wet conditions required by most plants, clay pots can also become saturated and produce unattractive algae or slime on the outside.

The size of the container is usually unimportant. However, most carnivorous plants may be grown in half-pots, that is, pots in which the height is approximately two thirds of the maximum width. These have two distinct advantages: they are economical because of the smaller amount of growing medium they hold; and they need less water because of their reduced volume, which can be helpful during prolonged dry spells, especially where water

The mature root system of a *Pinguicula* plant

storage facilities are limited. The relatively small pot size does not matter, since the root systems of most carnivorous plants take up remarkably little space. (For the few exceptions, where larger pots are required, see pp.36, 40 and 41).

WATERING

Watering is generally very simple. Each pot should stand in a plastic saucer, about 1 in. (2.5 cm) deep and at least 1 in. (2.5 cm) wider than the pot, which is kept almost full of water throughout the growing season. In the cooler conditions of winter, when the majority of plants experience a period of dormancy, the soil should remain just damp. They may then require water once or twice a month. However, the few tropical species which continue in growth, such as tender *Pinguicula*, should be stood in a saucer of water all year.

Watering from above is usually best avoided. It can initiate rotting, especially where moss is growing round the plant, and water droplets left on the leaves may cause scorch, particularly with sticky-leaved plants. In most cases, the wide saucer full of water provides sufficient humidity and misting or syringing is not required. Nor is it advisable to fill the pitchers of Sarracenia and similar plants with water. If the soil, water and humidity are satisfactory, the pitchers will control their own water content.

On the whole, rainwater or other soft water is the most suitable for carnivorous plants. The majority of them grow in the wild in acid or neutral soils and will not tolerate lime, and certainly few plants appreciate the chlorides and fluorides added to household tap water. Rainwater, usually plentiful in Britain, is easy to collect in barrels fed by downpipes from house or greenhouse roofs. Water from a local river or lake, so long as it is acid or neutral, is equally good. Expensive alternatives, which are feasible only for the smallest collections, are distilled and deionized waters. Distilled water is absolutely safe. However, prolonged use of deionized water (which has been treated to replace certain chemicals, including the unwanted calcium, with others) could lead to a build up of chemicals in the soil; to avoid any risk of damage, the soil should be changed within six months.

In the event of lime-free water being temporarily unobtainable, the simplest solution is to use household tap water, which in the short term will harm very few plants. If at all possible, it should be first boiled and cooled before use. In hard water areas, tap water should be resorted to for the minimum amount of time and, if this does unavoidably run to several weeks, the compost should be changed as soon as regular supplies of rainwater become available again.

There are some plants which are lime-tolerant and may be given tap water, again preferably boiled and cooled before use. They include many of the hardy Pinguicula and the tropical Nepenthes.

TEMPERATURES

There are carnivorous plants to suit all the various temperature ranges found in nature and mimicked in greenhouses (see table 4). Although most gardeners are surprised to learn that any of them are hardy, there is indeed a choice of plants for growing outside in temperate climates. Many species which are not considered hardy will also survive long periods of frost, but because growth is severely set back, there is little point in growing them outdoors. Perhaps surprisingly, most of the hardy species will grow as well under glass as when fully exposed (except during the coldest winters when some protection may prove beneficial). Normally, protection only serves to advance the growth period. However, hardy species of Pinguicula, like P. grandiflora, produce fewer off-

Table 4: minimum winter temperatures for carnivorous plants

Temperature range	Suitable plants
outdoors (hardy)	*Sarracenia purpurea purpurea,* *S. flava; Drosera rotundifolia,* *D. anglica, D. filiformis filiformis;* *Pinguicula grandiflora, P. vulgaris;* *Utricularia vulgaris, U. australis,* *U. intermedia, U. minor*
32°F (0°C) minimum	all *Sarracenia; Drosera binata,* *D. filiformis traceyi, D. peltata,* *D. whittakeri; Dionaea muscipula;* *Pinguicula grandiflora; Darlingtonia* *californica; Utricularia praelonga,* *U. intermedia*
50°F (10°C) minimum	all tropical *Pinguicula,* especially *P. agnata, P. gypsicola* and the *P. moranensis* group; most *Utricularia;* tuberous, pygmy and tropical *Drosera,* especially *D. pygmaea, D. aliciae, D. capensis* and *D. adelae;* highland *Nepenthes;* all *Sarracenia* except *S. purpurea* *purpurea; Cephalotus follicularis*
70°F (21°C) minimum	lowland *Nepenthes*
64°F (18°C) minimum (days) 52°F (11°C) minimum (nights)	highland *Nepenthes*

sets when given winter protection, while *P. vulgaris* definitely prefers to be grown outdoors and attempts to maintain it in a greenhouse are likely to fail.

For a 32°F (0°C) winter minimum, some of the Australian tuberous *Drosera* such as *D. peltata* and *D. whittakeri* are worth experimenting with. Although the plants rest as tubers for about eight months of the year, they add tremendous variety to the genus and are quite easy to grow, even tolerating short periods of sub-zero temperatures.

All *Sarracenia* do well when cool and frost-protected, especially *S. purpurea purpurea* and *S. flava,* which do not like warmth in winter. Some *Pinguicula* are also hardier than expected and certainly many of the *P. moranensis* group will survive for a few days at 32 to 50°F (0–10°C), with the exception of *P. moranensis* *caudata.*

17

Above: the tuberous sundew, *Drosera whittakeri*, a rosetted plant with large white flowers

Below: *Nepenthes ampullaria*, a small lowland species which is suitable for a terrarium

For the avid enthusiast, the 50°F (10°C) minimum will allow the greatest range of carnivorous plants to be grown. Many of the hardier plants, including Sarracenia, will still perform well in these temperatures and all the tender ones will survive and grow happily, apart from a few of the truly tropical species. This is the safest temperature at which to maintain new plants if their preferences are unknown. Highland Nepenthes, previously thought to need higher temperatures, also succeed in these conditions, providing the humidity is very high.

Among plants which are readily available at present, only the lowland Nepenthes require higher temperatures than these.

LIGHTING

Two factors will contribute to the health of your plants – light intensity and the number of hours of light received out of 24 hours, referred to as day length.

Most carnivorous plants do best in a position which receives the maximum amount of light and too little will cause them to grow long and spindly, making very poor specimens. In the home they may be placed in a south-facing window, or a north-facing one in the case of Pinguicula. In the greenhouse, some precaution should be taken against scorching from the sun. The toughest of plants, for example Sarracenia, will tend to burn around the trap lip, especially where nectar droplets magnify the rays of the sun. Light shading in summer is quite sufficient to protect them and several materials are available from garden centres for the purpose. A tinted or white plastic sheet or a single layer of coarse-meshed netting can be fixed inside the greenhouse roof or windows, or a proprietary shading paint can be applied to the glass. These methods will also do for Drosera (apart from the pygmy and tuberous species, which appreciate any degree of strong sunlight).

Among the more specialist plants, Utricularia present a different problem. Although many withstand direct sunlight, too much will promote the growth of moss and algae, which quickly smother the leaves and may kill the plants. Nepenthes are far more sensitive and prefer to be grown in shade. They can be considered to be similar to many of the tropical orchids.

Pinguicula need some care. The thousands of leaf glands produce a coating of mucus, which invariably leads to damage from scorching, and it is advisable to allow no direct sunlight to fall on them. This has the advantage that the leaves often develop a generalized pink to rose hue or reddish veining. These plants show their true splendour only when grown out of direct sunlight, in shade or, for best results, under artificial lighting.

Artificial lighting can be used with most carnivorous plants and is particularly suitable for *Pinguicula*, *Utricularia* and *Nepenthes*. It eliminates the risk of scorching and guarantees that light will be available for the required time each day, irrespective of weather conditions. On the other hand, the lighting must normally be placed near the plants, which detracts from their beauty, and it can be expensive to run.

Fluorescent tubes designed for plants (such as Grolux) are cheap and easy to obtain. They remain cold and so can be placed very close to the plants. This is essential, as the light they emit is not effective with carnivorous plants at distances beyond about 1 to 2 ft (30–60 cm). For very small or rosetted plants, the distance should be 6 in. to 1 ft (15–30 cm). However, large upright plants will be difficult to light in this way and *Sarracenia* often become limp and straggly, since the base tends to be shaded by the upright pitchers.

For large collections, tall plants and *Nepenthes*, there are more powerful light sources, which not only light a small room with a single bulb, but also radiate heat and help to maintain the required temperature range. These systems (for instance, Sunlighter) can be bought through advertisements in plant magazines. All artificial lighting is best controlled by a compatible automatic switch or clock. During the growing season, usually in summer, provide nine to twelve hours of light; in winter provide seven hours.

Day length is important to most carnivorous plants. It may influence flowering and can trigger or interrupt growth in those which have a natural dormancy period. A summer day will give eight or more hours of light and in these conditions, plants produce their summer foliage and may flower. With shorter days, many plants become dormant. This is particularly true of the hardy sundews, *Drosera*, while the pygmy sundews enter a reproductive stage at the same time. Short days also cause *Sarracenia* and *Cephalotus* to produce only non-carnivorous leaves and other plants, such as *Darlingtonia*, cease growth until the days lengthen.

Generally speaking, nature should be left to its own devices. Only when a plant fails to flower or stubbornly refuses to end its dormancy should alteration of day length be considered.

PROPAGATION

Various propagation techniques work for carnivorous plants, but whatever the method used, new plants are best potted on only

Pinguicula rosei, showing the fine leaf colour which develops out of direct light

after a root system can be seen. If plants are transplanted before this occurs, it is necessary to maintain high humidity, keep them well watered and ensure that they are warm and receiving plenty of light. Direct sunlight should be avoided for very young plants, especially those with no root system. It is usually advisable to begin propagating in spring or early summer to produce strong plants in time for the winter.

Seed

Almost all the plants available will set viable seed, with a few notable and annoying exceptions, for instance, the various pygmy *Drosera* and *Utricularia*. Seed can also be purchased from various sources, including some commercial suppliers and specialist plant societies.

All seed can be sown on the surface of the growing medium suitable for mature plants (see under the descriptions, pp.32–55) and germinated in the light, but out of direct sunlight. Most seed will germinate at 55 to 65°F (12.5–18°C) and need not have any bottom heat or other special treatment. *Nepenthes* require higher temperatures and higher humidity.

Germination can take from two to three weeks, to as much as two years with some tuberous sundews. As long as seedlings are well protected from direct sunlight, all can be treated as adult

21

plants. Seed of some carnivorous plants will germinate only if fresh and it is generally advisable to sow seed as soon as possible after collection.

Seed of hardy plants should be exposed to frost first and will germinate in spring as temperatures climb above 50°F (10°C). In most cases the seeds can be sprinkled on to a bed or tray of sphagnum, placed outdoors and covered with a sheet of glass to prevent them being washed away by heavy rain. If sown in a bog garden, the seed should be protected from birds with chicken wire.

Cuttings

The horizontal stems or rhizomes of *Sarracenia* may be cut into sections about 1 in. (2.5 cm) long, with a clean sharp knife, taking care not to damage the roots. Each section is then planted and treated as an adult plant. Make sure it lies horizontally with any large roots pointing downwards and do not allow soil to cover the top of the rhizome. Take rhizome cuttings in spring, during pitcher growth. The remainder of the original plant should be left with several good roots. *Sarracenia* can rot if allowed to form thick clumps and rhizome cuttings should be taken every three or four years.

Several other plants respond well to the taking of root cuttings, including *Drosera*, especially those with thick fleshy roots such as *D. capensis*, and *Cephalotus*. The method is basically the same as for *Sarracenia*, except that the cut sections should be covered to their own depth with soil. Stand the pot in water to keep the soil moist and, when the leaves appear, the plants can be repotted and treated as adults.

Leaf cuttings are suitable for a variety of plants. With *Drosera*, entire leaves are removed from the stem, making sure that the leaf base, where it attaches to the stem, is also taken, and laid sticky side up on a bed of growing medium, shredded live sphagnum moss or perlite. The medium should be watered and then a layer of live shredded sphagnum moss placed over the leaf and sprayed lightly. The pot or container is covered with a piece of glass, a clear plastic lid or clear plastic film to maintain humidity and placed in a light warm position, but out of direct sunlight. Watering or spraying will be required until new plants are seen to emerge from the leaf cuttings. These can be safely potted on when a root system has developed. (If these daughter plants are removed at a very early stage, the original leaf cutting will often continue to produce new plants. As the daughters are still relatively fragile, they should be kept warm, humid and out of direct

Propagation of *Pinguicula* from leaf cuttings placed on perlite

sunlight until a root system has developed.) Use only the healthiest-looking leaves and ensure maximum contact with the growing medium for best results, by shaping the soil to fit the leaves or by carefully pinning them down with thin wire. Check daily for any movement of the leaves and reposition them if necessary. Different species of *Drosera* react in different ways and it is worth experimenting, by cutting the leaves into sections or using parts of the leaf blade rather than the stalk.

Leaf cuttings of *Dionaea* can be treated in exactly the same way, although it is essential that the entire leaf is taken as new plants will develop from a bud at the base. Use new leaves as soon as they mature and discard the traps, which often move and inter-fere with development of the new plants.

The leaves of tender species of *Pinguicula* are ideal for leaf cut-tings. Although summer trapping leaves can be used, most growers prefer to take the dormant winter leaves, which are more plentiful, are not sticky and will not detract from the beauty of the plant in summer by their absence. Up to three quarters of the total number of winter leaves on each plant can be safely removed. Place these on the surface of the growing medium or perlite. (Perlite has the distinct advantage of being free of moss spores, which allows the new plants to grow without competition.) Water the soil to ensure it is damp, cover the container to maintain humidity and keep it in a warm light place out of direct sunlight. After a few weeks, the leaves will begin to show the growth of new plants as tiny buds. They should be potted on as soon as a root system has started to develop or when they are beginning to over-crowd each other. To increase the yield, simply cut each leaf into smaller sections before placing it on the medium. New plants grow from any point of damage. As if this method were not simple enough, some species, such as *P. primuliflora*, actually grow daughter plants on their leaf ends without assistance. These new plants can be left to root or potted on as soon as they are big enough to show leaves. As with young plants from *Drosera* leaf cuttings, maintain a high humidity for your *Pinguicula* plants.

Nepenthes can be propagated from stem cuttings. Take off the top of the plant as a length bearing three or four pairs of leaves and remove the base pair. Cut off half of each remaining leaf. Dip the cut end of the stem in rooting hormone and plant this in live sphagnum. The plant and pot should then be wrapped in a clear plastic bag and kept warm, as for adult plants, checking regularly that the sphagnum does not dry out.

Gemmae

Gemmae are highly modified leaves designed to protect the plant during dormancy or to increase natural propagation. The most dramatic example is found in the pygmy *Drosera*, which during dormancy produce an abundance of gemmae from the centre of the plant. If left unchecked, they will often smother and kill the parent plant and a toothpick or small blunt piece of plastic can be used to dislodge them. Keep the plant away from others when doing this, as the gemmae part from the parent at some speed and will grow where they land! Collected gemmae can be sown as seeds, with the small dent in the surface facing up. The growing bud develops from the base of this dent. Gemmae can be stored for several weeks on damp paper, sealed in plastic and kept in a refrigerator.

Division

Darlingtonia and *Cephalotus* can be increased by dividing be-tween clumps on the rhizome. Always ensure that divisions are

24

free from decay by cutting out any tissue which is brown rather than white.

Other plants divide naturally and can slowly be increased in this way. Tuberous species like *Drosera whittakeri* may produce new underground daughter tubers, but these are best left untouched for one or two years. Stoloniferous species, including *Drosera prolifera* and most *Utricularia*, will spread laterally and can be simply divided from the parent with a knife if the growth remains permanent. Hardy *Pinguicula* will produce offsets or daughter buds during the winter, which can be removed from the base of the parent in late winter and treated as mature plants.

FERTILIZERS

All carnivorous plants are adapted to catch their own food because they live in areas where the soil lacks nutrients. If they are planted in soil that contains much in the way of chemical nutrient, one of two things usually happens. At best, the plants are oversupplied with food and do not form traps. Such plants are of little value to the collector. At worst, the chemical concentration (even though weak for normal plants) is so high that it damages the roots and kills the plant. Similarly, foliar sprays can injure the plant through overfeeding.

However, there are some cases when the very careful use of a weak foliar feed can be beneficial. Seedlings and immature plants can be encouraged to grow by applying a foliar feed suitable for houseplants (such as Baby Bio), at about one quarter the recommended concentration. Plants should be sprayed no more than once a month and always in the early morning or evening to avoid any risk of leaf burn. This treatment can be given to *Sarracenia* and *Darlingtonia* seedlings, *Nepenthes* and *Pinguicula* (applying very light mistings only).

Growing outdoors

HARDY PLANTS

It is still not widely known among gardeners that some carnivorous plants are perfectly hardy and tolerant of even severe winters. In fact, all the three most popular groups have representatives which are hardy to one degree or another (see table 4, p.17), requiring habitats varying from bogs to standing water.

For perfect conditions, a boggy area or pond is needed. Those gardeners who already have a natural site merely have to choose which plants to grow. Otherwise, an artificial bog or pond should be constructed. To do this, first mark out the chosen area, preferably with a length of garden hose, which is easier to work with than other materials. Most ponds and bogs benefit from a surround of stone slabs, so allow for this too. Remove soil and any turf within the marked area to a depth of 12 to 18 in. (30–45 cm), ensuring that the bottom is level and that the sides are cut at an angle of about 45°, to prevent them collapsing. For a bog this single excavation is sufficient. For a pond, mark out another area within the hole, leaving a margin of at least 12 in. (30 cm) width. Dig out this area again in the same way.

Next, check that there are no large stones in the hole. To protect the liner from future damage by hidden stones, put a 1-in. (2.5 cm) layer of any available sand over the flat surfaces. For the stone slab surround, remove turf and soil to the required depth from the perimeter and then add a layer of sand. A pond liner, several varieties of which are available from garden centres, must now be fitted. To calculate the size of liner required, use the maximum dimensions of the pond or bog and the following formula: [length + twice maximum depth] × [width + twice maximum depth]. For example, if the pond is 10 ft long and 8 ft wide, with a maximum depth of 2 ft, the liner would measure: $[10 + (2 \times 2)] \times [8 + (2 \times 2)] = 14 \times 12$ ft.

To fit the liner, place it over the hole and secure it at the corners with stones. Pour soft water into the centre until the liner is weighed down and remove the stones to allow the liner to settle on the sand. To produce a pond, continue to fill the remaining space with more soft water. For a bog, cut a $\frac{1}{8}$-in. (3 mm) drainage hole at a depth of 6 in. (15 cm) in the middle of each long side of the liner. Then almost fill the bog with a well mixed combination of damp or wet moss peat and washed sharp sand, approximately

Utricularia intermedia, one of the few hardy aquatic bladderworts, may be grown in a shallow pond or peat bog

5 parts moss peat to one part sand. (Sand can be omitted but helps the water penetrate the mixture.) Saturate the bog with soft water before adding the top layer, a 2-in. (5 cm) deep mixture of washed sharp sand, wet moss peat and live chopped sphagnum moss. This final layer should then be well watered, again using soft water. The bog or pond can now be completed by covering the edge of the pond liner with stone to form a surround.

Incidentally, a bog garden is an excellent alternative to an ordinary pond kept full of water, which poses a hazard to young children. It is also far more attractive than the common replacement for the pond – a sand pit. The artificial bog is still an unusual feature in gardens, but it allows experimentation with a new range of plants.

Carnivorous plants should not be exposed to the risk of burning by wind and midday sun. However, sensible planting of other bog-loving plants will provide shelter, while adding to the natural look of the bog. These could include medium-sized plants such as the flag iris, *Iris kaempferi*, and marsh marigold, *Caltha palustris* (for further recommendations, see the Wisley handbook, *Water gardens*). It is also a good idea to group the carnivorous plants together, interspersed with other plants which are better at binding the soil with their roots. Otherwise, birds will find the damp soil an irresistible attraction and spend many glorious hours uprooting the plants in their search for grubs, worms and other food. For small areas, an efficient but unsightly alternative is to cover the bog in chicken wire mesh or net, which can be supported on a frame or short stakes above the surface. If preferred, the wire can be laid on or just below the surface and holes cut for the initial planting. Although the wire will eventually rot, the area will by then have a good covering of moss and most plants will be well enough established to withstand the assaults of birds.

Planting in the bog is simply a question of digging a small hole and inserting the rather small root system of the plant. As the soil is wet, no watering in will be required. In a pond the plants can be grown in pots, which are placed in the water in the same way as with other aquatic or marginal plants. Remember that the acid water of such a pond is not suitable for fish.

Only the hardiest carnivorous plants will survive the winter in or under water, although many will do so in the bog garden. These can be moved to the pond between spring and autumn or planted permanently in the bog. Sarracenia, for example, can be grown with the crowns either submerged or exposed and will accept temporary flooding lasting as much as several months. The pond is ideal for Utricularia, providing a permanent home for the hardy species; U. australis will even tolerate some lime. (See table 5).

Table 5: outdoor habitats suitable for hardy carnivorous plants

Habitat	Plants
Rock garden	Pinguicula grandiflora (lime-tolerant)
Scree (very well drained)	Drosera peltata
Bog	Sarracenia purpurea purpurea, S. flava; Drosera rotundifolia, D. anglica, D. filiformis filiformis; Pinguicula grandiflora, P. vulgaris
Pond margin	all hardy plants suitable for bogs, especially Pinguicula grandiflora
Pond (true aquatics)	Utricularia vulgaris, U. australis, U. intermedia, U. minor

Pinguicula grandiflora is particularly good for the edge of a pond

ALPINE PLANTS

Although many of the true alpines among carnivorous plants are rare, there are a few more readily available species which should succeed in a rock garden or scree.

There is often a natural hollow at the foot of the slope of a rock garden, which acts as a sump, draining the surrounding land, and as a consequence is always damp. Some of the alpine *Pinguicula*, especially *P. grandiflora*, are invaluable for such a spot and beautiful even without their attractive flowers.

On a really well drained scree slope the tuberous sundew, *Drosera peltata*, is well worth a trial. The soil should be almost pure gravel or grit mixed with only a small proportion of moss peat. The plant will need shelter from wind and will not want to be cold and wet, particularly in winter, so a cover may be required. It should do well in a situation which suits *Soldanella*. However, prolonged cold spells in wet conditions will damage the tubers and it is advisable to keep some spare plants indoors or under glass.

Growing indoors

A number of carnivorous plants make good house plants (see table 6). A position near a south-facing window suits the majority of them. The tall *Sarracenia* species in particular require full sun and will otherwise grow into long thin pitchers which are unable to support themselves. All pitcher plants need maximum light to develop their colour. However, there is some risk of damage from the rays of the sun which, when magnified by glass, may cause leaf scorch. Plants with fleshy leaves, especially *Pinguicula*, are most susceptible and should not be placed in direct sunlight. Those *Sarracenia* which produce large amounts of nectar on their pitcher lid, like *S. flava*, may also suffer slight leaf scorch unless protected against the strength of the midday sun.

Table 6: carnivorous plants suitable for growing indoors

Position	Plants
Windowsill in direct light	*Sarracenia flava* – dislikes too much winter heat. *S. purpurea venosa* – horizontal pitchers, easy. *S. × catesbaei* – beautiful hybrid of the above two. *Drosera capensis* – easy, grows and flowers quickly. *D. spatulata* – easy, small rosette. *D. aliciae* – easy, red hairs on green leaves. *Cephalotus follicularis* – unusual appearance. *Dionaea muscipula* – easy, trap colour red in sun.
Windowsill in indirect light	All tender *Pinguicula* – good flowers, dislike cold. Most *Utricularia* – small and fascinating. *U. sandersoni* – almost always in flower, easy.
Shade	*Pinguicula moranensis* forms – *P. caudata* and *P. mexicana*.

The Albany or west Australian pitcher plant, *Cephalotus follicularis*, may be grown in a sunny window

Many *Pinguicula* thrive on a north-facing windowsill or one that receives only indirect light. They can be protected from short periods of direct sunlight by placing them behind ordinary household net curtains. Some of them, including the commonest forms of *P. moranensis* (usually sold under the names *P. caudata* and *P. mexicana*) can be grown in shade, away from a window.

Carnivorous plants grown indoors require no special treatment. They can be grown in a pot of the appropriate compost and placed in a plastic saucer containing 1 in. (2.5 cm) of rain water or distilled water. During dormancy the plants should not stand in water but should be watered once a week to keep the compost damp. Tender plants, especially *Pinguicula*, should be moved away from a window when outside temperatures fall below 50°F (10°C).

Popular carnivorous plants

Of the more than 1,600 carnivorous plant species and hybrids so far identified, many are hard to obtain or difficult to grow. This chapter therefore concentrates on those plants which are most suitable for the beginner, all of them being easy to grow, inexpensive, readily available and fascinating or beautiful. (The compost types referred to at the end of the cultivation notes are described in table 3, p.13).

DIONAEA VENUS FLY TRAP

Perhaps the best known carnivorous plant is the Venus fly trap, *Dionaea muscipula*. This incredible little plant is often responsible for arousing interest in carnivorous plants as a group and yet has given rise to most of the misconceptions associated with them.

The sole representative of the genus, it comes from the Carolinas in eastern North America. The plant grows the green trap-bearing leaves, approximately 5 in. (12.5 cm) long, for most of the year, except in winter. The extraordinary trap mechanism consists of two slightly fleshy pads, hinged together at the end of the leaf, each with a series of fierce-looking but soft spines arranged round the edge, almost resembling eyelashes. The inner surface of the pad may be flushed pink or, in good specimens, an intense red and bears three small erect 'hairs', which act as triggers. As a precaution against accidental triggering of the trap by falling debris or rain, either two separate hairs must be touched or one must be touched twice; in addition, if these two movements do not occur within a specific time period of about 2 to 20 seconds, then the trap will remain open.

Once the trap is sprung, the leaf goes through several amazing changes. In the first stage, the pads bend in towards each other rapidly, so that the fringes of soft spines interlock. Any further action must be stimulated by the efforts of the victim to escape, as another protection against mistaken identity. If it is indeed an insect, the trap now enters the second stage and closes tight, suffocating or crushing it. This sealing of the trap is the result of cells growing on the outside of the pads – the fastest growth known in any plant. Enzymes are then emptied on to the prey, which is

Opposite: the Venus fly trap is famous for its dramatic trap action

The fringe of *Dionaea*, as well as the trap pads, may be deep red when grown in direct sunlight

slowly dissolved, nutrients are absorbed through the trap pads and a short burst of leaf production usually follows. Each trap can be re-used three or four times before it dies naturally and is replaced.

Dionaea produces insignificant white flowers in late spring, held high above the traps to enable insects to pollinate without being caught. Seed production can be copious, but should not affect the vigour of mature healthy plants.

Cultivation

A windowsill in direct sunlight is an excellent place for *Dionaea*, although it will survive happily in a cold greenhouse. Grow it in a half-pot and stand in a tray or saucer of soft water for most of the year to keep the compost wet. During winter, when *Dionaea* is dormant, keep the compost merely damp by watering occasionally. Some success has been recorded with plants outdoors, but if overwintered outside, they never grow to their best and will be killed in severe weather. Compost: A. Temperature: 50°F (10°C) winter minimum for best results.

Propagation

Seed germinates easily on shredded sphagnum moss or pure perlite, preferably in a temperature of 70-80°F (21-26°C).

SARRACENIA NORTH AMERICAN OR TRUMPET PITCHER

The pitcher plants of North America are probably the most popular of the carnivorous plants. Although only a handful of species exist, there are several subspecies or varieties of each and an almost bewildering choice of hybrids. They range in height from the prostrate forms barely a few inches tall, with horizontal pitchers sometimes 8 to 12 in. (20-30 cm) long, to the largest of the upright forms which can be a majestic 3 ft (1 m) high. Pitcher colours also vary, from light yellow through green to intense purple, often overlaid with splashes of deeper colour on the throat or veining. Perhaps the queen of this eye-catching genus is S. *leucophylla*; the lower part of the pitcher is green, which gradually fades to near or pure white at the top, with a network of deep red veins.

The flowers precede the pitchers from early spring to mid-summer and deservedly attract attention. They are held proudly aloft at or above the normal full height of the pitchers and come in colours from cream through yellow to brick-red as well as gorgeous dusky pinks and oranges. When the four large petals fall, the central part of the flower, a massive umbrella-shaped style, is left in all its glory until middle or late summer. These flowers are also good for cutting or drying and, although a few have a slightly unpleasant odour, others have a delightful perfume. Pitcher production continues into late summer. Finally, some species develop strap-shaped winter leaves to take the plants through into spring.

The trap itself is a complicated structure and manages to catch its prey without movement. (*Sarracenia* are therefore passive traps, contrary to popular opinion.) At the pitcher head is a large fixed lid, which acts as a landing platform and is often marked with veins to lead flies down into the throat. It may also be generously supplied with nectar glands to attract the prey. Short hairs make it easier for the insect to maintain a foothold here but, as it descends, downward-pointing hairs decrease the opportunity for retreat. Further down into the trap, it will reach a more slippery area, where nectar is produced in abundance, and then well into the throat, a smooth wide band, which is capable of secreting digestive enzymes. On this slippery surface the fly will inevitably lose its footing and fall to the bottom of the pitcher, while the downward-pointing hairs on the walls prevent any

Sarracenia flava is a reliable and attractive trumpet pitcher

escape. At least the intoxicating nature of the nectar must make death quite painless.

Cultivation

Grow *Sarracenia* in good light, but avoid direct sunlight through glass, which can cause leaf burn. Use full-length pots for all except the prostrate plants and stand in a tray of soft water. In winter, however, the compost should be kept damp, not wet. Prolonged winter temperatures above 60°F (15°C) interrupt the normal period of dormancy and often result in poor specimens the following year. Compost: A. Temperature: 32°F (0°C) winter minimum; *S. purpurea purpurea* is hardy.

Propagation

By rhizome cuttings of mature plants or from seed. Seeds may be presoaked to aid germination, which otherwise usually begins after about eight weeks but may take several years.

Recommended plants

S. *flava*: easiest, up to 3 ft (1 m) tall, flowers and pitchers bright yellow.
S. *purpurea purpurea*: horizontal or semi-erect purple pitchers, very robust (see p.63).
S. *leucophylla*: upright, majestic, white cap, red veined (see p.13).
S. × *catesbaei*: natural hybrid, easy, yellow with brick red flowers.

DROSERA SUNDEW

Although many countries are home to at least one species of *Drosera*, it is surprising that the sundews are so rare in plant collections. They offer a wide variety in both size and shape and, as well as the leaves, which are spectacularly beautiful in sunshine, many have the bonus of attractive flowers.

Plants consist mainly of one type of leaf, adapted for its role as an insect trap. The flattened leaf surface is covered in fine hairs ending in a series of glands, which secrete a clear, colourless, sticky fluid or mucus to form a dew drop on each hair. Insects seem attracted to this or possibly to the red leaf colour found in some species, but once they touch the fluid, their fate is almost certainly sealed. The dew is remarkably sticky and rarely lets go of its prey and, in its struggles to escape, the insect gradually becomes covered in glue until it eventually drowns. Many sundews accelerate the process of capture by bending the leaf over to form something akin to a 'fly sandwich'. After trapping, all *Drosera* bend the tiny hairs towards the prey and then secrete digestive juices on to it, absorbing nutrient through the leaf. The dead carcase often remains stuck to the leaf, but in the wild may be blown away.

The leaves of *Drosera* are generally green, with the hairs or tentacles contrasted in red. The dew drops add to the effect by turning into mini-rainbows whenever they are struck by sunlight. Flowers are produced reliably by most species and although individually lasting for only a day in some cases, they are often large and showy or borne in profusion to provide a display for weeks or even months. The flower colour varies, with only blue being absent.

Sundews are normally grouped as follows, a system based on their appearance or country of origin rather than on botanical distinctions.

Pygmy sundews

Limited to Australia, these small sundews are frequently about
¾ in. (2 cm) across, rarely up to about 1¼ in. (3 cm). All are rosette-
forming, but a few grow vertical stems with the old leaves
remaining attached. Although small, the flowers can be as large as
or larger than the plant. They open only in strong light, often for
just one or two hours.

Cultivation

Most pygmy sundews will survive in a cold greenhouse.
However, winter temperatures should be kept slightly higher,
preferably above 50°F (10°C). In summer, provide the maximum
amount of available light. Poor light may kill the plants and over-
wintering under fluorescent tubes is beneficial. These *Drosera* are
unique within the genus in reproducing from small specialized
leaves called gemmae, each reduced to a tiny scale and having a
dormant bud at the base. Gemmae are produced when light is
available for less than eight hours per day – in other words, winter
in the northern hemisphere, corresponding to a period of
dormancy for normal leaf growth. Stand half-pots in soft water
for most of the year, keeping the compost damp, not wet, in
winter. Compost: A. Temperature: 50°F (10°C) winter minimum
for best results.

Propagation

Best from gemmae, treating these as seed. Leaf cuttings can also
be used with some species.

Fork-leaved sundews

This superb and easily grown group consists of several varieties
or forms of one species, *D. binata*. As a rule, the longer the name
the more branched the variety will be, *D. binata multifida extrema*
having leaves with over ten forks. The species itself has leaf stems
up to 12 in. (30 cm) long, bearing a single Y-shaped fork, which is
covered in a profusion of tentacles. Flowers are always numerous
and either white, pink or pink-edged.

Cultivation

The fork-leaved sundews are happy if kept frost-free, but some
varieties will remain in growth all year if warmer.

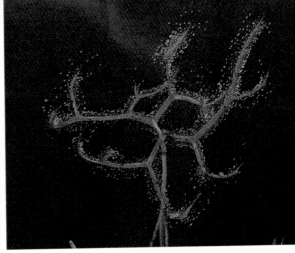

Left: the flower of *Drosera capensis*, a South African sundew

Right: a fork-leaved sundew, *Drosera binata multifida*, with many branches to each leaf

Below: *Drosera capensis* makes a pretty, easily grown houseplant

Propagation

Simple from thick root cuttings. Leaf cuttings can also be taken. *D. binata* (the species, not the varieties) is easily raised from seed. Provide good light and stand half-pots in a tray of soft water while in growth, keeping the compost damp, not wet, during winter. Compost: A. Temperature: 32°F (0°C) winter minimum; 50°F (10°C) minimum enables *D. binata multifida* to remain in growth all year.

South African sundews

These include the most easily grown carnivorous plants and fortunately some of the most beautiful, both rosette- and stem-forming. Most rosetted plants will be up to 2 in. (5 cm) in diameter, the other types growing between 6 and 12 in. (15-30 cm) tall. They have attractive leaves and colourful or large flowers.

Cultivation

These make good windowsill plants and the easier species, including *D. aliciae* and *D. capensis*, remain decorative and in growth all year. Plants in half-pots should stand in a tray of soft water and be kept warm all year, in good light.

The large-rooted *D. regia* and *D. cistiflora* benefit from being grown in deeper pots than usual. Both are susceptible to over-watering, especially *D. cistiflora*, which lies dormant for most of the year. Full-length pots should be used, as the larger amount of growing medium will retain moisture longer between periods of careful watering. When in growth, both plants can be grown in pots standing in trays of soft water; when growth is slow or dormancy begins, the compost should be only just damp. Compost: A. Temperature: 50°F (10°C) winter minimum.

Propagation

By seed and leaf cuttings. *D. capensis* will also develop well from thick root cuttings.

Tuberous sundews

This is another predominantly Australian group of sundews and one of the most interesting. The plants die down to rest as tubers during their native hot dry season and, although seldom above ground for more than two or three months, they are well worth growing.

There are three distinct forms of tuberous sundew. Rosetted plants, like D. whittakeri, consist of large leaves with few but large tentacles, the whole being up to $1\frac{1}{2}$ in. (4 cm) across. Upright plants, for instance D. peltata, have one or more stems and many taller species appear to rely on each other or other plants for support. They have numerous leaves on obvious stalks, the leaves often being circular, with the top removed as if bitten out to create a three-quarter-moon shape. These appear all the more bizarre because of the relatively large number of tentacles, which glisten beautifully in sunshine, giving rise to the common name, rainbow sundew. The third type, known as fan-leaved sundews, comprises only three species, which occur mostly in humid tropical areas. D. stolonifera belongs to this group.

Cultivation

In the northen hemisphere, these sundews make active growth in winter. Water is essential at that time, but can cause the tubers to rot if they are too wet or conditions are too cold. On the other hand, many species also require drying out during the dormant period in summer, although if this occurs too soon, the new tubers will fail to develop. To overcome these problems, tuberous sundews are best grown in full-length pots 12 in. (30 cm) deep, filled with river or sharp sand mixed with a very little moss peat or leaf litter. The greater volume of growing medium will help to control water content, while the depth of sand will allow the tubers to dry off naturally when dormancy begins.

Keep the plants warm all year, with a minimum of 50°F (10°C) in winter. Stand the pots in water or water from above when the plants are growing and stop giving water as they die down. Several species, particularly D. whittakeri and D. peltata, can be kept wet all year and will survive frost-free temperatures. However, if experimenting with growing these in lower temperatures, it is advisable to have the soil damp rather than wet. The majority of tuberous sundews require strong light, but D. stolonifera should not receive direct sunlight through glass. Compost: A, but with very little peat. Temperature: 50°F (10°C) minimum; D. peltata and D. whittakeri will tolerate lower temperatures if protected from excessive moisture.

Propagation

By seed, which can take several years to germinate.

Left: the unique leaf shape of the tuberous sundews, in this case *Drosera peltata auriculata*

Right: a red-flowered form of the normally white-flowered *Drosera adelae*

Hardy northern sundews

These species develop tight winter-resting buds and are tolerant of cold, although not all are reliably hardy in extreme winters. The British native, *D. rotundifolia*, is among the smallest of this group, forming a rosette of about 1 in. (2.5 cm) diameter, with tentacled pads on longer leaf stems. The North American *D. filiformis* is very different, having huge wire-like leaves over 8 in. (20 cm) long, held upright in strong sunlight and entirely smothered in tentacles.

Cultivation

All these sundews must be cool in winter, many preferring to be outdoors. They like good light, but not direct sunlight through glass. Plants kept outside should be in permanently wet compost. For those protected from frost, the compost should be damp in winter. *D. filiformis tracyi* is not completely hardy, while *D. filiformis filiformis* is. Compost: A or D. Temperature 32°F (0°C) minimum for *D. filiformis tracyi*; most others prefer exposure to winter frost.

Propagation

Best from seed.

Recommended plants

D. pygmaea: easy pygmy sundew, increases quickly.

D. binata: beautiful, forked leaves, windowsill plant (see p.39).

D. capensis: easy windowsill plant, leaves bend to trap (see pp.13 and 39).

D. aliciae: very easy rosette, suitable for windowsill.

D. peltata: upright plant, easiest of the tuberous types.

D. whittakeri: easy rosetted tuberous form, large white flowers (see p.18).

D. stolonifera: easy tuberous, many forms, large white flowers.

D. adelae: must be kept warm, large, leaves well coloured in sun.

D. prolifera: stoloniferous, spreads quickly, keep warm and humid.

D. rotundifolia: hardy, small rosettes, for the peat bog (see pp.8 and 63).

D. anglica: hardy, largest of the British sundews.

PINGUICULA BUTTERWORT

Well known to many orchid growers are the butterworts. Their sticky leaves are particularly efficient at attracting and killing the small flies that attack orchids and can be used as a most effective living fly-paper.

Apart from their usefulness, these are some of the most attractive plants available. The leaves form fleshy rosettes 1 to 9 in. (2.5–22.5 cm) in diameter, of a delightful shade of light green, and the surfaces are covered in the greatest concentration of dew-producing tentacles found in any carnivorous plant, giving an effect unique in the plant world. Added to this, the flowers are not only large and showy but, in some tender species, are borne twice a year. They make excellent plants for the garden or the home.

All *Pinguicula* use their leaves as lures and traps, but movement is limited to the leaves curling at the edges, which prevents digestive juices running off. When an insect is trapped on the leaf margin, the leaf sometimes rolls over it to increase the surface area in contact with the victim and so improve digestion. This also protects against any possible leaching effect of the rain. The sticky trap works in the same way as a sundew leaf (see p.37), but appears to be better at catching small flies.

The leaves are produced in spring and almost all the species die back into some form of dormancy in autumn. Hardy plants such as *P. grandiflora* and *P. vulgaris* develop resting buds to resist the severest weather, while tender ones generally form a tight group of reduced winter leaves, creating a charming winter rosette.

Above: *Pinguicula moranensis mexicana*, like others of its group, is an excellent indoor plant

Below: a butterwort in the winter resting stage; summer leaves are beginning to emerge from the middle

Cultivation

Hardy species do far better if kept outdoors in winter. If over protected, they are likely to be slightly stunted and will reproduce more slowly. They will also be damaged by high summer temperatures and must be grown outside to ensure survival, or at least in a very well ventilated cool greenhouse. With plants at the edge of a pond or in a bog, ensure that the compost remains damp or wet during growth; in a greenhouse, stand the pots in water all year.

Tender species, such as any of the *P. moranensis* group (good varieties of which include *P. moranensis caudata* and *P. moranensis mexicana*), need warmth in winter, not just frost protection. A minimum of 50°F (10°C) is suitable, except for *P. primuliflora* which requires higher temperatures and humidity. In summer, their succulent or fleshy leaves should be protected from strong sun and good indirect light can reveal the subtle colours. For all tender species the compost should be permanently moist, with the pots standing in a tray of water. Compost: A, with a little leaf mould added, for most; D, for *P. vulgaris*; A or B with lime added, for *P. grandiflora*. Temperature: 50°F (10°C) minimum for tender species; cool in summer and exposure to frost for true hardy species.

Propagation

Increase hardy species by seed, which must be sown fresh, or by offsets. For tender species, leaf cuttings are the most reliable method. Mature plants will also produce divisions naturally during the winter, which can easily be separated by hand and grown on as adults. With *P. gypsicola*, the leaf cuttings are treated slightly differently. The winter leaves are used and, after removing them from the plant, they should be allowed to dry out for a day before placing on the compost. If the leaves are left even longer to develop new plant shoots, seen as tiny buds at the cut end, excellent results can be obtained.

Recommended plants

P. grandiflora: hardy, lime-tolerant, rockery or pond edge, easy (see pp.29 and 64).
P. vulgaris: hardy, purple flowers, peat bog, not easy.
P. moranensis group: large, easy indoors in indirect light or shade (see pp.6 and 62).
P. agnata: large, indoors, white flowers tinged purple at edge.
P. gypsicola: needs care, very beautiful leaves and flowers.
P. primuliflora: keep warm, leaves grow new plants at ends (see p.46).

The sinuous leaves of *Pinguicula alfredae* give it a bizarre appearance; unfortunately, it is not easy to obtain

DARLINGTONIA COBRA LILY

A relative of *Sarracenia*, *D. californica* is another pitcher plant from North America and the sole representative of its genus. As the name implies, it is a native of California, where it grows in the mountains along the mossy banks of streams.

With its startling cobra-like appearance, which gives it the common name of cobra lily, this plant certainly deserves to be better known to most gardeners. The pitcher trap works in a very similar way to *Sarracenia*, but looks different, the entire pitcher length being twisted and the head forming an expanded hood with snake-like fangs at the mouth. Nectar on the hood and fangs attracts the prey, which enters the hood through a hole. Although it might still be able to escape at this stage, it is usually lured further into the hood and away from the true exit, towards the light admitted by transparent cells or false windows. Exhaustion or sheer bad luck leads the fly on into the slippery pitcher neck, from where there is almost no hope of escape.

The mottled green traps may continue to be attractive and possibly useful for two to three years. Each season the new traps tend to be bigger than those of preceding years, the very first pitcher of the year being the largest. As with *Sarracenia*, the flowers appear before the pitchers in spring and are tall pendulous lanterns of greenish yellow and brick-red. Although the petals fall within a few weeks, the seed heads remain decorative for months and also dry well.

Cultivation

In the wild, *Darlingtonia* survives snow cover, but Californian winters are relatively mild and dry. Therefore, it is wise to provide winter protection, either under glass or outside with a covering of loose straw. In summer, a cool root run is essential and plants should be kept in semi-shade outdoors, where they will happily grow in a bog or pond. Alternatively, they may be potted and stood in water, preferably submerging the roots and crown and in winter reducing the depth of water to about 2 in. (5 cm). Compost: D or B. Temperature: 32°F (0°C) winter minimum or protect with straw.

Propagation

New plants produced around the edges of the pot can simply be separated and repotted.

CEPHALOTUS ALBANY OR WEST AUSTRALIAN PITCHER PLANT

The third type of pitcher plant, *C. follicularis*, is the only member of its genus and is confined to a small area of western Australia near the town of Albany.

To see this little plant, never more than $2\frac{1}{2}$ in. (6 cm) tall, is to want to own one. It nestles in a bed of moss, forming a tight clump of open-mouthed pitchers, green in normal light or purple in strong sunshine, with leaves which are more or less numerous depending on the season. Each pitcher is like a miniature, thumbless, boxing glove, with a cap overhanging the opening and three obvious hairy ribs on the main body, as well as a smoothly ribbed rim. These ribs almost certainly help to guide insects, chiefly ants and beetles, into the trap. Nectar is also secreted as a lure. As with *Darlingtonia*, the trap is spotted with windows which admit light, although their exact purpose or even whether they have a role in trapping is not clear. Whatever the explanation, *Cephalotus* is a strange, lovable, but much under-studied plant. Traps are produced in the summer months and plain, slightly fleshy leaves in spring. The flower spikes are an incredible 2 ft (60 cm) tall, but of no particular beauty.

Cultivation

The plant does best in a very open compost and care should be taken that it is not overgrown by tall mosses. It should stand in a tray or deeper container of soft water all year. Strong sunlight is essential for the best growth and colouration. Compost: B. Temperature: 50°F (10°C) winter minimum (see pp.31 and 59).

Propagation

Easily increased by division, root cuttings or leaf cuttings.

NEPENTHES MONKEY CUP

Probably the most familiar pitcher plants of all are the tropical *Nepenthes*, known as monkey cups. Although few people have seen them alive, they are popular subjects for magazines and television and in Victorian times were common as hothouse plants.

   Like all the pitcher plants, the *Nepenthes* have a pitfall or passive trap, into which insects fall to drown in a soup of digestive juices. The pitcher is designed to ensure that the insect is attracted to a landing platform, the lid; it is then led on to the obviously ribbed rim and, once over this, to the inner waxy surface, where it tumbles into the waiting pool of liquid (to join not only other past meals but a host of animals, including mosquitoes and tree frogs, which use the pitchers as homes in the wild).

   The traps hang on tendrils from the ends of beautiful, green, glossy leaves and are truly pitcher- or jug-shaped. They come in an enormous range of colours, shapes and sizes for, as well as a large number of species, there are innumerable hybrids, many of them created by the Victorians.

   Pitcher production is at its peak in summer and occurs even on seedlings. Many plants have two types of pitcher, one at ground level and the other further up the vine-like stem, which enhances their appeal. Pitchers are from 2 to 14 in. (5-35 cm) long and the vines may grow to 50 ft (15 m), but do best if cut back annually.

Opposite: one of the many *Nepenthes* hybrids, with their spectacular pitchers and attractive leaves

The trap of *Nepenthes ampullaria*, with its lid or landing platform

Cultivation

The Victorian practice of growing *Nepenthes* in hothouses led to a belief that all need tropical temperatures. Fortunately, this is not true. Many are highland species, requiring cooler humid conditions to recreate their cloud forest habitat, and these will be happy if the nights are frost-free and the days cool or warm. Only the lowland species need tropical conditions, in other words, warm nights and hot days (see table 7).

Table 7: winter and summer temperatures suitable for lowland and highland *Nepenthes*

	Winter		Summer	
	Day	**Night**	**Day**	**Night**
Lowland species	70°F (21°C) minimum		as high as possible	
Highland species	64–71°F (18–22°C)	52°F (11°C) minimum	71–95°F (22–35°C)	as cool as possible, but 52°F (11°C) minimum

The traps of monkey cups are found in a wide variety of shapes and sizes

All *Nepenthes* dislike sunlight and, although their tolerances vary, most prefer about 50 per cent shading. Plastic sheets, netting or shading paint can be used. Netting has the advantage of being easily removed in winter, when less shading is required, and also does not inhibit ventilation too much. Pitcher development is in fact influenced by humidity rather than light. Too little light will affect only the colour of the pitcher and it is therefore better to overshade and then reduce the shading gradually, checking carefully for any sign of leaf or pitcher burn.

Regular misting or spraying three or four times a day is essential and the humidity can also be raised by damping down the greenhouse floor or placing plants above a large tray of water. An automatic misting device can be a worthwhile investment. *Nepenthes* will happily accept tap water, although in hard water areas the lime may form deposits that stain the leaves. Good drainage is vital and is easily provided by growing the plants in baskets. If this is not possible, they can be grown in pots of pure live sphagnum (in which case soft water rather than tap water must be used, in order not to kill the moss), with a live sphagnum wick to draw water up from a reservoir into the pot. Unlike most carnivorous plants, *Nepenthes* should be watered from above and must not stand in water, otherwise the roots will rot and the plants soon die. Compost: C or D. Temperature: see table 7.

Propagation

Readily increased from stem cuttings and, as they become leggy with age and benefit from cutting back in spring, this should provide an ample supply of cuttings. They can also be raised from seed, which must be sown fresh on shredded sphagnum or pure perlite and kept at a temperature of 80°F (26°C). Mist regularly and do not allow the growing medium to dry out. Germination will occur in two to six weeks. Seedlings burn easily and should be well protected from direct sunlight.

Recommended lowland plants

N. ampullaria: small red and green traps on small terrestrial vine (see pp.18 and 50).
N. gracilis: small, easy, green or spotted forms.
N. rafflesiana: large, vigorous, easy, beautiful beginner's plant.

Recommended highland plants

N. khasiana: common, easy, good beginner's plant.
N. tentaculata: very small, straggler, beautiful spotted pitchers.
N. alata: small, easy, some forms with subtle shades.

UTRICULARIA BLADDERWORT

Although widely distributed throughout the world, this large genus of carnivorous plants is virtually unknown except to specialists. There are probably several reasons: the plants and often the flowers are rather small; they are always found in very wet areas or in water and are often quite difficult to locate, their size again not helping here; and the process of capturing prey is only visible with the aid of specialized photography and, even when the action is slowed, is completed in a fraction of a second.

The traps, unlike those of the majority of carnivorous plants, are not obviously modified leaves but small sacs produced on the thin stems or stolons that make up the bulk of the plant. These bladders are kept relatively empty of water and have a trap door with an accompanying trigger hair. The slightest movement of the trigger causes the door to snap open and water is then free to rush in and fill the partial vacuum in the bladder, carrying with it any passing creature, including hopefully the creature that triggered the reaction. The trap door then shuts and the bladder becomes a stomach into which digestive juices are emptied. Even if it already contains a meal, the trap can work again as soon as the internal pressure is reduced by pumping water out.

The tiny flowers of *Utricularia* are often described as orchid-like

The plant usually consists of minute leaves attached to a creeping stem or stolon, although in a few species, notably the South American tree-dwelling ones, the leaves are up to 6 in. (15 cm) long. The small flowers, often produced unpredictably, are delightful and fascinating. Bladderworts do not have roots, the function of absorbing food from the soil having been replaced by an efficient trapping process. They grow only where water is permanently or seasonally abundant. Terrestrial species inhabit bogs or wet savannah; epiphytic (tree-dwelling) species obtain sufficient moisture from the humid air of tropical rain or cloud forests and the occasional downpour of torrential rain; and aquatic species float on or grow under the water. Where the supply of water is seasonal, they either survive the drought as tubers or grow as annuals.

Cultivation

Of the aquatic *Utricularia*, only a few are hardy. They all succumb to attack by fish and snails and become quickly overgrown with algae unless precautions are taken. Either they should be grown in very strongly acid conditions, in which algae, fish and snails cannot survive, or sufficient plant cover, such as water lilies and duckweed, should be provided to prevent algae growth. Resting buds are formed at the bottom of the pond to overwinter.

Utricularia sandersoni, a tropical non-aquatic bladderwort with abundant flowers

U. intermedia and several other species enjoy very shallow water up to 1 in. (2.5 cm) deep and will creep out on to wet peat. Most non-aquatic species should be kept permanently wet, using shallow pots or trays of compost and standing these in water all year. *U. reniformis*, a large South American species, is liable to rot if kept wet, but must not be allowed to dry out. The commonly available species, apart from the hardy ones, will generally grow well if temperatures are maintained above 50°F (10°C) all year. Compost: A, for *U. sandersoni* and most non-aquatic species; D, for *U. longifolia*, *U. praelonga*, *U. calycifida*; B, for *U. reniformis*. Temperature: 50°F (10°C) minimum for tender species; cool summer and exposure to frost for hardy species.

Propagation

By division; even the smallest piece should grow well if kept moist. Some grow from seed easily, but a few such as *U. subulata* and *U. bisquamata* (*U. capensis*) are fairly invasive and for this reason should be isolated from others.

Recommended plants

U. intermedia: yellow flowers, aquatic to creep over peat bed, hardy (see p.27).
U. sandersoni: most reliable, small, pale lavender flowers, tiny leaves.
U. reniformis: large kidney-shaped leaf, orchid-like purple flower.
U. praelonga: semi-aquatic, yellow flowers twine up support.

Utricularia menziesii from southwest Australia is unusual in being tuberous and has remarkable flowers

Pests and diseases

Carnivorous plants are luckily troubled by very few pests and diseases and the grower should be able to control any problems relatively easily.

BOTRYTIS (grey mould)

As with all diseases, prevention is better than cure. Only one disease is likely to cause any real trouble, this being the fungal infection *Botrytis*. An outbreak is easy to spot, the fungus appearing as a grey fuzz of very slim hairs and, if mature, creating a cloud of grey dust when disturbed. The dust consists of spores, which will spread the infection to other plants, so disturbance must be kept to a minimum and treatment should involve the entire growing area when a major attack occurs. The fungus enjoys cool humid conditions and quickly colonizes dead plant tissue, providing there is sufficient humidity. In winter the risk of an outbreak can be reduced by keeping all soil damp rather than wet. Congested growing plants or plants harbouring dead material, which has fallen between tightly packed leaves or remained attached, are most susceptible and this applies particularly to *Sarracenia*. To avoid problems, dead or unattractive leaves should be pulled off, making sufficient space for air to circulate. Pitchers or winter leaves of *Sarracenia* can be pulled carefully and will peel away from the rhizome to leave a leaf scar. Dead flower stems should be removed in the same way.

In the event of an outbreak, any fungicide available in retail packs can be used at the recommended strength for other plants. Bad attacks, which usually occur in winter in poorly ventilated greenhouses, should be treated with a tecnazene smoke, obtainable from most garden centres.

APHIDS (greenfly)

Even carnivorous plants are subject to these annoying pests and their relatives. Although very few mature plants are attacked, immature growth is vulnerable. In the case of *Sarracenia*, the trap tops are likely to become deformed, while with *Pinguicula*, which suffer most of all, the new summer leaves can be badly weakened unless immediate action is taken. Once the summer leaves have unfolded, however, attacks are rare even on successive leaves.

The dew-tipped tentacles of a sundew leaf can be used to remove
aphids from other carnivorous plants

Sarracenia psittacina has green leaves coloured red or purple according to light intensity

Pesticides, such as pirimicarb and malathion, can be used, but not too frequently. An alternative is to mop up the offending insects with the sticky leaf of a sundew, especially *Drosera capensis*. Sundews appear unaffected even by major infestations.

SCALE INSECTS

Although very unusual with carnivorous plants, these pests can occasionally attack *Nepenthes*, *Sarracenia* and *Darlingtonia*. The brown, almost circular, and sometimes elongated scales are rarely seen to move, but they are both unattractive and damaging, weakening the plant by sap-sucking and quickly reproducing themselves.

Most proprietary pesticides intended for scale insects are very strong, because of their resistance to many chemicals. It is therefore recommended that the pests are removed with tweezers – a time-consuming job unless done at the first sign of infestation. Alternatively, they can be pushed off with cotton wool, dipped in methylated spirits, on the end of an orange stick. *Sarracenia* and *Nepenthes* should be able to withstand treatment with malathion, but care should be taken to avoid large amounts of chemical entering the traps.

Very little else should afflict carnivorous plants. Red spider mite sometimes appears and can be treated with malathion. If a plant fails without any obvious sign of damage, inspect the roots and soil for signs of root borers and remove them by hand. Dead or infected tissue should be cut out with a sharp knife.

The intriguing design of *Cephalotus* makes escape virtually impossible

Sources of supply

GARDEN CENTRES AND NURSERIES

As carnivorous plants gain in popularity, they can be found at an increasing number of garden centres. For the beginner, a garden centre could be a good starting point, giving an opportunity to see the plants available and to examine them before purchasing.

Specialist nurseries stock a much greater range of carnivorous plants, including both the easy and popular ones and the rarer species and hybrids sought by keen collectors. Most of these nurseries cater for visitors, either individuals or groups, although it is always best to make an appointment. Most of them also offer a mail order service, which is a convenient and reliable method of obtaining plants. Reputable suppliers will always avoid sending plants that would be damaged in the post and will replace any that arrive in unsatisfactory condition.

When choosing carnivorous plants, look for strong healthy specimens. The leaves should show no sign of wilting and should be free from pests and fungus. Pitchers should be upright, while fly-paper traps should appear sticky. Many carnivorous plants, including *Dionaea* and tropical *Pinguicula*, will normally have one or two dead or dying leaves, but these should never predominate. Seedlings and very young plants are easily damaged by underwatering or sudden changes in temperature and are best avoided when starting a collection. Any plants purchased from garden centres should be repotted as soon as possible in the recommended potting compost (see table 3, p.13).

SPECIALIST SOCIETIES

Carnivorous plants may also be obtained by joining a specialist society. There are several carnivorous plant societies throughout the world and one in Britain, which provides a seed and plant exchange scheme for members. In addition, the society gives advice, publishes a newsletter and journal, exhibits at flower shows, holds lectures and meetings and arranges field trips.

Pinguicula x *mola*, a hybrid of *P. moranensis caudata* and *P. gypsicola*

Addresses

(Price lists are usually free, but the inclusion of a stamped addressed envelope is always appreciated.)

Cyril G. Brown
65 Highfield Crescent
Hornchurch
Essex RM12 6PC

Heldon Nurseries
Asbourne Rd
Spath
Uttoxeter ST14 5AD

Marston Exotics
Hurst Lodge
Martock
Somerset TA12 6JU
(correspondence address only)

Sarracenia Nurseries
Links Side
Courtland Avenue
Mill Hill
London NW7 3BG

W. T. Neale & Co Ltd
16/18 Franklin Rd
Worthing
Sussex BN13 2PQ

Carnivorous Plant Society
174 Baldwins Lane
Croxley Green
Herts WD3 3LQ

Opposite: butterworts of the *Pinguicula moranensis* group are some of the most attractive carnivorous plants available

Above: The eye-catching umbrella-like flowers of *Sarracenia purpurea purpurea*

Below: The diminutive *Drosera rotundifolia* is a British native

The hardy *Pinguicula grandiflora* is easy to grow outside